# EMMANUEL JOSEPH

# HOW ORGANIZATIONS CAN HELP CLIMATE CHANGE

# Contents

1

# Chapter One: INTRODUCTION

C limate change is a pressing issue that is affecting our planet in profound ways. As individuals and communities, we have a responsibility to take action and make positive changes to mitigate the impact of climate change. One crucial way in which we can address this global issue is through organizational efforts. Organized collective action by businesses, governments, and institutions can have a significant impact on reducing greenhouse gas emissions, promoting sustainability, and fostering environmental stewardship. In this essay, we will explore how organizations can play a vital role in combating climate change and creating a more sustainable future for generations to come.

Before furthering into knowing how organization can help climate change, we need to understand the two variables; organization and climate change.

## 1. WHAT IS AN ORGANIZATION?

An organization is a group of people who come together to achieve a common goal or purpose. This can include businesses, non-profit organizations, government agencies, and other groups that work together to accomplish specific objectives.

## 1. WHAT IS CLIMATE CHANGE?

Climate change refers to the long-term alteration of temperature and typical weather patterns in a place. It is primarily caused by human activities, such as burning fossil fuels, deforestation, and industrial processes, which release greenhouse gases into the atmosphere. These gases trap heat and cause the Earth's temperature to rise, leading to a range of environmental impacts, including rising sea levels, extreme weather events, and disruptions to ecosystems

## 1. WAYS IN WHICH ORGANIZATION CAN HELP CLIMATE CHANGE

Due to climate change outcomes, many companies are and will be in the future, greatly affected in many different ways. Yet, many do not know what they can do to help fight it . However, organizations play a crucial role in addressing climate change because they have the power to influence and implement sustainable practices on a large scale. By reducing their carbon footprint, implementing energy-efficient measures, and supporting green initiatives, organizations can help mitigate the effects of climate change and contribute to a more sustainable future.

Therefore, the ways in which organization can help climate change includes;
4.1.MEASURING AND ANALYZING GREEN GAS EMISSIONS:

Greenhouse gas emissions are gases that trap heat in the Earth's atmosphere, leading to the global warming phenomenon and contributing to climate change. The main greenhouse gases include carbon dioxide ($CO_2$), methane ($CH_4$), nitrous oxide ($N_2O$), and fluorinated gases.

To help combat climate change, organizations can play a crucial role by measuring and analyzing their greenhouse gas emissions. This process involves identifying and quantifying the amount of each greenhouse gas the organization is responsible for emitting, typically measured in metric tons of $CO_2$-equivalent.

One way to measure and analyze greenhouse gas emissions is by conducting a comprehensive greenhouse gas inventory. This inventory will document all the sources and amounts of greenhouse gas emissions associated with the organization's operations. It may include direct emissions from on-site activities, such as fuel combustion in buildings and vehicles, as well as indirect emissions from purchased electricity and business travel.

After completing the inventory, the organization can use this data to set emission reduction targets, develop strategies to reduce emissions, and track progress over time. By understanding their carbon footprint, organizations can make informed decisions to

lessen their impact on the environment and contribute to global efforts in combating climate change.

Moreover, by publicly reporting their greenhouse gas emissions and reduction initiatives through platforms like the Carbon Disclosure Project (CDP) or the Greenhouse Gas Protocol, organizations can demonstrate transparency and accountability in their commitment to sustainability.

Basically, measuring and analyzing greenhouse gas emissions is a crucial step for organizations to take responsibility for their environmental impact, reduce their carbon footprint, and contribute to mitigating climate change. By implementing sustainable practices and embracing renewable energy sources, organizations can make a positive difference in the fight against global warming.

2

# Chapter Two: REDUCING ENERGY CONSUMPTION:

Reducing energy consumption is a key strategy for organizations to mitigate climate change and promote sustainability. By implementing energy efficiency measures and practices, organizations can not only reduce their carbon footprint but also save costs in the long run. Here are some detailed ways in which organizations can help combat climate change by reducing energy consumption .

1. Conduct an Energy Audit: Start by conducting an energy audit to identify areas of. inefficiency and opportunities for energy savings. This involves analyzing energy use patterns, evaluating equipment performance, and identifying potential energy-saving measures.

2. Invest in Energy-Efficient Technologies: Upgrade to energy-efficient appliances, lighting systems, heating, ventilation, and air conditioning (HVAC) systems, and other equipment. Energy-efficient technologies can significantly decrease energy consumption and lower utility bills.

3. Implement Smart Building Controls: Use smart building technologies, such as energy management systems and smart thermostats, to optimize energy usage based on occupancy, weather conditions, and operational needs. This can help reduce energy waste and improve overall efficiency.

4. Establish Energy Conservation Policies: Develop and implement energy conservation policies and employee awareness programs to promote a culture of energy efficiency within the organization. Encourage employees to turn off lights, computers, and other equipment when not in use.

1. Utilize Renewable Energy Sources: Transition to renewable energy sources, such as solar, wind, or hydroelectric power, to reduce reliance on fossil fuels and lower greenhouse gas emissions. Consider installing solar panels or purchasing renewable energy credits to support clean energy production.

4.2.6 Optimize Transportation Practices: Encourage carpooling, telecommuting, and the use of public transportation among employees to reduce transportation-related emissions. Provide incentives for using fuel-efficient vehicles or electric cars.

1. Implement Sustainable Practices: Adopt sustainable practices such as recycling, composting, and reducing waste to minimize the environmental impact of operations. Consider sourcing materials locally and using eco-friendly products to support sustainability efforts.
2. Monitor and Track Energy Performance*: Regularly monitor and track energy usage through energy management software or tools. Set energy reduction targets, track progress, and make necessary adjustments to achieve goals effectively.

Therefore ,by taking proactive steps to reduce energy consumption, organizations can not only contribute to combating climate change but also improve their bottom line through cost savings, enhance their reputation as environmentally responsible entities, and contribute to a more sustainable future for all.

3

# Chapter Three: REDUCING WASTE AND FIGHTING OBSOLESCENCE:

Reducing waste and fighting obsolescence are essential strategies for organizations to combat climate change and promote sustainability. By implementing waste reduction initiatives and adopting practices to extend the lifespan of products and resources, organizations can minimize environmental impact and contribute to a circular economy. Here are some detailed ways in which organizations can help fight climate change by reducing waste and obsolescence:

1. *Implement a Waste Reduction Plan: Develop a comprehensive waste reduction plan that includes strategies for waste prevention, recycling, composting, and diversion from landfills. Conduct a waste audit to identify sources of waste and opportunities for improvement.
2. Promote Product Lifecycle Management: Embrace a product lifecycle management approach that considers the entire lifecycle of a product, from design and production to use and disposal. This can help optimize resource efficiency, reduce waste generation, and enhance product quality and durability.

1. Embrace Circular Economy Principles: Embrace circular economy principles by designing products for longevity, reusability, and recyclability. Implement take-back programs, product refurbishment, and remanufacturing initiatives to extend product lifespan and reduce waste.

2. Reduce Packaging Waste: Optimize packaging design to minimize waste generation and use eco-friendly materials. Implement packaging reduction strategies, such as right-sizing packaging and using recyclable or biodegradable materials, to reduce environmental impact.

3. Encourage Repair and Reuse: Promote repair and reuse practices by offering repair services, refurbishing products, and facilitating product exchange or sharing among customers. Encourage a culture of reuse and resourcefulness within the organization and among stakeholders.

4. Adopt Sustainable Procurement Practices: Source materials and products from suppliers that prioritize sustainability, ethical practices, and waste reduction. Consider factors such as product durability, recyclability, and end-of-life disposal when making procurement decisions.

5. Educate and Engage Stakeholders: Educate employees, customers, and suppliers about the importance of waste reduction, circular economy principles, and sustainable practices. Encourage stakeholders to adopt environmentally friendly behaviors and participate in waste reduction initiatives.

6. Monitor Performance and Set Targets*: Monitor waste generation, diversion, and reduction efforts through waste tracking systems and performance indicators. Set measurable targets for waste reduction, track progress regularly, and adjust strategies as needed to achieve goals effectively.

By reducing waste and fighting obsolescence, organizations can not only help combat climate change but also reduce resource depletion, minimize pollution, and create economic opportunities through resource efficiency and innovation. Embracing a circular economy mindset can drive positive change by transforming the way products are designed, produced, used, and disposed of.

Additionally, organizations can collaborate with industry partners, policy-makers, and community stakeholders to exchange best practices, advocate for sustainable policies, and support initiatives that promote waste reduction and circularity on a broader scale. By working together, organizations can amplify their impact and contribute to a more sustainable future for all.

incorporating waste reduction strategies, promoting product longevity, and embracing circular economy principles are key steps that organizations can take to address climate

change and environmental challenges. By prioritizing sustainability and embracing innovative solutions, organizations can make a meaningful contribution to building a more resilient and resource-efficient economy while safeguarding the planet for future generations.

# 4

# Chapter Four: OPTIMIZING EMPLOYEES:

Optimizing employees' transportation can have a significant impact on reducing carbon emissions and combating climate change. Here are some ways organizations can help:

1. Encouraging Sustainable Commuting Options: Organizations can encourage employees to use sustainable modes of transportation such as walking, cycling, carpooling, and public transportation. Providing incentives like flexible work hours, parking discounts for carpoolers, or subsidizing public transportation passes can motivate employees to choose greener options.

2. Setting Up Telecommuting Programs: Allowing employees to work from home a few days a week can reduce the need for daily commuting, leading to lower carbon emissions. Remote work also promotes work-life balance and saves employees time and money on transportation.

3. Providing Facilities for Biking: Installing bike racks, showers, and changing rooms can encourage employees to bike to work. Offering bike storage or bike-sharing programs can make it easier for employees

to choose cycling as a sustainable commuting option.

4. Implementing Electric Vehicle (EV) Charging Stations: Installing EV charging stations in the workplace can encourage employees to switch to electric vehicles, reducing greenhouse gas emissions from traditional gasoline-powered cars.

5. Educating and Raising Awareness: Organizations can organize workshops, seminars, and campaigns to educate employees about the environmental impact of transportation choices and raise awareness about sustainable commuting options.

By implementing these strategies and promoting a culture of sustainability within the organization, employers can help reduce their carbon footprint, contribute to combating climate change, and create a more environmentally friendly workplace.

# 5

# Chapter Five: MOBILIZE FOR THE CLIMATE CHANGE CHALLENGES

Mobilizing for climate change challenges is crucial for organizations to address the urgent need for sustainability and environmental stewardship. Here are some ways organizations can help:

1. Setting Clear Sustainability Goals: Organizations can establish measurable sustainability goals related to reducing carbon emissions, increasing energy efficiency, minimizing waste, and promoting sustainable practices throughout the company's operations. By setting clear targets, organizations can track progress and hold themselves accountable for their environmental impact.

2. Implementing Green Initiatives: Organizations can implement green initiatives such as reducing single-use plastics, adopting renewable energy sources, optimizing resource management, and promoting eco-friendly practices in day-to-day operations. These initiatives can help reduce the organization's carbon footprint and environmental impact.

3. Partnering with Environmental Organizations: Organizations can collaborate with environmental organizations, non-profits, and gov-

ernment agencies to support climate action initiatives, participate in advocacy campaigns, and contribute to conservation efforts. By joining forces with like-minded partners, organizations can amplify their impact and drive positive change on a larger scale.

4. Engaging Employees and Stakeholders: Organizations can engage employees, customers, suppliers, and other stakeholders in climate change awareness campaigns, sustainability workshops, and community events. By fostering a culture of environmental responsibility and encouraging collective action, organizations can mobilize a broader community to tackle climate change challenges together.

5. Investing in Sustainable Innovation: Organizations can invest in research and development of sustainable technologies, products, and services that help mitigate climate change and promote a circular economy. By embracing innovation and supporting green solutions, organizations can lead by example and inspire others to follow suit.

By mobilizing for climate change challenges, organizations can play a significant role in reducing carbon emissions, protecting the planet, and ensuring a sustainable future for generations to come.

## ORGANIZATIONS HELPING IN THE FIGHT AGAINST CLIMATE CHANGE LISTED BY FOODTANK

1. **350.org**, International

Author and activist Bill McKibben and a group of university friends founded 350.org in 2008, with the goal to keep the global carbon dioxide concentration under 350 parts per million. They are using the power of collective individuals internationally to stop oil and gas development and move to 100 percent renewable energy.

2. **Arab Forum for Environment and Development** (AFED), Arab Region

AFED's mission is to promote environmental education and provide space for environmentalists to come together. They advocate for sustainable development, and influence planners, decision makers, businessmen, civil society, and media. AFED hosts an annual Health and the Environment in

Arab Countries Conference and also publishes the AL-BIA WAL-TANMIA (Environment & Development) magazine.

3. **Asia Pacific Adaptation Network** (APAN), Asia and Pacific Region

To build more resilience against the consequences of climate change, APAN supports adaptation across Asia and the Pacific Region. APAN provides resources and tools for planning, financing, and building adaptive societies. For example, there are publications on Coastal Zone Management that can help coastal communities reduce damage from sea-level rise.

4. **Biomimicry Institute**, International

Biomimicry is a design technique that solves problems by mimicking nature. Biomimicry Institute's mission is to promote the transfer of ideas, designs, and strategies from biology to sustainable human systems design. For example, someone who wants to spend less energy building might consider using Moist Brick, a naturally cooling building material that can condense water from nighttime air similar to the skin of a Texas Horned Lizard.

5. **C40 Cities**, International

C40 brings together a network of megacities from around the world, allowing them to drive climate action through collaboration and knowledge sharing. New York City, Johannesburg, Hong Kong, Sydney, Tokyo, London, and Mexico City are just some of the cities on the list that have committed to the climate targets established in the Paris Agreement.

6. **Caribbean Community Climate Change Center** (CCCCC), Caribbean Region

The CCCCC works to coordinate the Caribbean region's response to climate change by identifying solutions. The Centre offers several tools and resources to help users take appropriate action in their communities. They also carry out a number of projects to address issues caused by climate change across the Caribbean region.

7. **Citizens' Climate Lobby**, International

As a 501(c)4 nonprofit organization, Citizens' Climate Lobby pushes for nonpartisan policies to address climate change. With over 600 local chapters internationally, Citizens' Climate Lobby builds political support for climate action by empowering individuals to use their own voice. They provide people with a toolkit to help with outreach, engagement, organizing, media, and lobbying.

8. **Climate Alliance**, International

Made up of municipalities and districts, regional governments, nongovernmental organizations (NGOs), and other organizations, Climate Alliance is one of the largest European city networks dedicated to climate action. The Alliance promotes actions to slow climate change in both European municipalities and the Amazon River basin.

9. **Climate Action Network** (CAN), International

CAN is a global network of more than 1,300 environmental NGOs. With regional hubs in regions including West Africa, South Asia, Latin America, and Eastern Europe, the Network works to promote governmental and individual action to address the impacts of climate change. CAN's working groups address a variety of issues including agriculture, science policy, and technology.

10. **Climate Cardinals**, International

With the belief that every person has the right to basic environmental education, Climate Cardinals' mission is to translate climate information into the native language of those who don't speak English. With the help of over 6,000 volunteers, this youth-led organization has worked in over 40 countries globally and helped 350,000 people understand climate change.

11. **Climate Collaborative**, United States

Focused on natural foods businesses concerned about climate change, the Climate Collaborative breaks climate action down into three simple steps: commit, act, impact. To date, 673 businesses have made commitments, and the Climate Collaborative offers a variety of resources and support to these businesses to help them reach their goals.

12. **Climate Group**, International

The Climate Group is working toward net-zero emissions by 2050 by holding organizations accountable for climate commitments they make. International business and government leaders come together to make commitments during Climate Week NYC. Then, Climate Group holds them accountable by sharing the actions they've taken and connecting them to other groups with similar objectives.

13. **Climate Justice Alliance** (CJA), United States

CJA brings together frontline, community-based organizations to lead the just transition from an extractive economy to a regenerative one. The 70 member communities are small grassroots organizations working locally to fight climate change with traditional ecological knowledge. CJA unites these members with each other to scale up their impact nationally.

14. **Earthjustice**, United States

Earthjustice uses the power of the law to protect communities' and the planet's health. With offices across the U.S., their work has helped to save wildlands, halt destructive logging and mining, and encourage more sustainable farming policies. Their sustainable food and farming program works to improve worker safety, promote climate-friendly farming practices, and reduce pollution by factory farms.

15. **Environmental Defense Fund** (EDF), United States

For over 50 years, EDF has brought together scientists and lawyers to protect the environment. Using strategic partnerships, scientific and economic research, and advocacy, EDF works to strengthen laws and policies that improve the environment and public health. To do so, EDF takes on a wide variety of cases that range from overfishing and food contaminants to pollution from the oil and gas industry.

16. **Environmental Working Group** (EWG), United States

EWG researches how consumer products impact human health and the environment. A wide range of experts at the EWG makes it easier for consumers to understand the environmental impact of the products they purchase. EWG also annually releases the "Clean Fifteen" and "Dirty Dozen," which list the produce with the least and most pesticide contamination respectively.

17. **Extinction Rebellion** (XR), International

XR is a nonpartisan movement that demands governments declare a climate emergency, reach net zero emissions by 2025, and involve citizens in decision making. They use non-violent direct action and civil disobedience to communicate the urgency of the climate crisis. Because of the decentralized leadership, anyone from anywhere in the world can organize XR actions as long as it abides by core principles and values.

18. **Fridays for Future** (FFF), International

Started in 2018, FFF is a global climate strike movement that demands urgent action from government leaders. They work to put pressure on policymakers to listen to scientific experts about the consequences of climate change, ensure climate justice, and keep the global temperature rise below 1.5 degrees C compared to pre-industrial levels. FFF also offers a number of online resources for those interested in getting involved with the movement.

19. **Friends of the Earth**, International

Friends of the Earth uses the collective voice of grassroots members to speak truth to power and advocate for living in harmony with nature. They've caught the attention of large corporations and government agencies worldwide to demand that the rules of our political and economic systems need to change if we are to combat the climate crisis.

20. **Gender CC**, International

Created as a result of the United Nations climate negotiations (UNFCCC), GenderCC acknowledges that women play an important role in fighting climate change. This global network of organizations, experts, and activists are working to integrate gender justice into climate justice through raising awareness and empowering women.

21. **Greenpeace**, International

Founded in 1971, Greenpeace is a global organization that uses peaceful protest and strategic communication to highlight environmental issues and promote solutions. Now in more than 50 countries, Greenpeace works to halt deforestation, protect ocean health, stop nuclear testing, and more. Through solutions rooted in social justice, they hope to help communities disproportionately impacted by climate change.

22. **Health and Environment Alliance** (HEAL), Europe

HEAL works to shape laws and policies that protect planetary and human health and raise awareness about the benefits of mitigating climate change. Their goal is to create a toxic-free, decarbonized, and climate-resilient future. With over 90 member organizations, HEAL represents 200 million people across the 53 countries in the European region.

23. **Indigenous Environmental Network** (IEN), International

As part of the Native environmental justice movement, IEN formed in the U.S. after tribal grassroots youth and Indigenous leaders gathered to discuss the environmental assaults on lands, waters, communities, and villages. Today it connects Indigenous communities nationally and globally to protect sacred sites and natural resources, and support a just transition, carbon pricing, and the Green New Deal.

24. **Julie's Bicycle,** International

Julie's Bicycle supports the global Creative Climate Movement, helping artists use their creativity to become climate activists.  In addition to contributing to low-carbon creative programs, initiatives, campaigns, and communications, Julie's Bicycle developed The Creative Industry Green Tools, a set of free online carbon calculators. These calculators allow creative productions to measure their environmental impact, such as energy use and waste.

25. **La Via Campesina**, International

A grassroots network of more than 180 international organizations and 200 million farmers, La Via Campesina, fights for food sovereignty and better management of the world's resources. The group promotes agroecological farming techniques that work with the earth and help to mitigate climate change.

26. **Natural Resources Defense Council** (NRDC), International

With simple online actions anyone can take, plus three million members, and an international staff of experts, NRDC safeguards people, plants, animals, and natural systems.  By making strong partnerships across the United States, Canada, China, India, and Latin America, NRDC is pushing for climate solutions like solar power, electric vehicles, and national limits on carbon emissions.

27. **Naturefriends International** (NFI), International

Founded in 1895, NFI is one of the largest NGOs in the world.  With 350,000 active members, NFI advocates for environmentally and socially just tourism, and protects natural and cultural heritage sites. They provide activities and materials for experiencing nature and climate justice, such as an informative quiz about sustainable tourism.

28. **Oceanic Global**, International

Oceans store carbon and are integral in the fight against climate change. That is why Oceanic Global combines grassroots initiatives with industry solutions to shed light on humanity's essential relationship to the ocean. Through regional hubs in New York, Hamptons, Los Angeles, London, and Barcelona, Oceanic Global offers educational programming and community partnerships. The Oceanic Standard is their tool to help industries find sustainable vendors and improve their operations to keep the ocean healthy.

29. **Our Kids' Climate**, International

Originally founded in Sweden, Our Kids' Climate is a global network of parents who want to protect their kids from the climate crisis. Any group of parents from around the world can join the network to participate in family art projects and speak to mentors.

30. **Project Drawdown**, International

"Drawdown" refers to the point at which greenhouse gas emissions start to decline. Project Drawdown is an open-source and expert-reviewed resource that policymakers, universities, corporations, and activists around the world can turn to for climate solutions. For example, someone working in agriculture can learn how nutrient management techniques will impact their costs and reduce their carbon footprint.

31. **Rainforest Action Network** (RAN), United States

RAN works to make big corporations take responsibility for upholding human rights, keeping forests healthy, and protecting biodiversity. Their climate and energy team urges the financial sector to defund the extraction of fossil fuels and works to eliminate the forms of energy that create the most pollution: tar sands, coal mining and power, and liquified natural gas.

32. **South Durban Community Environmental Alliance** (SDCEA), South Africa

This group of 19 affiliate organizations advocates for clean air, clean water, healthy soil, and environmental justice in Durban, South Africa. Durban is a model city of sustainable development and SDCEA regularly liaises with the community, provincial and local government, industry, and commerce to promote a healthy, safe, and sustainable environment.

### 33. **Sunrise Movement**, United States

Sunrise is a youth movement to stop climate change and create millions of good jobs. One of Sunrise's current campaigns is to elect politicians who support the Green New Deal. In a recent Food Talk Live interview, Mackenzie Feldman, an organizer with Sunrise Movement Bay Area Chapter, says it's been inspirational to, "learn how much power we have as young people, and how we can push the political agenda and get things like the Green New Deal on the table."

### 34. **Union of Concerned Scientists**, United States

The Union of Concerned Scientists is a nonprofit with a mission to solve the planet's problems using science. Their team of 250 scientists, analysts, policy, and communication experts report the latest findings in the areas of climate, energy, transportation, food, nuclear weapons, and democracy. The Union of Concerned Scientists also fights disinformation and explains how special interests intentionally spread misleading information on climate change.

### 35. **World Wildlife Fund** (WWF), International

WWF is an international nonprofit that helps local communities access cutting-edge conservation science to protect natural resources. Local WWF chapters all around the world are tackling climate change by preparing for potential future disasters, and studying how these changes will impact ecosystems and wildlife.

### 36. **Zimbabwe Small Holder Organic Farmers' Forum** (ZIMSOFF), Zimbabwe

ZIMSOFF works to improve the lives of smallholder farmers across Zimbabwe who are practicing sustainable agriculture. With over 19,000 members and four regional clusters, they fight for food sovereignty, land justice, and environmental justice at local, national, regional, and international levels. Through the support of agroecology, organic agriculture, and open-pollinated varieties, they hope to support more environmentally friendly agricultural practices that will help mitigate climate change.

## CONCLUSION

In conclusion, the role of organizations in addressing climate change cannot be overstated. By implementing sustainable practices, adopting green technologies, and collaborating with stakeholders, organizations can make a meaningful

difference in reducing carbon emissions, conserving natural resources, and promoting environmental sustainability. As individuals, we can support and advocate for organizations that prioritize climate action and sustainability. Together, we can work towards a greener, more resilient planet for current and future generations. It is through collective effort and commitment that we can combat climate change and create a more sustainable and equitable world for all.